AMAZING MILITARY MACHINES

MILITARY HELICOPTERS

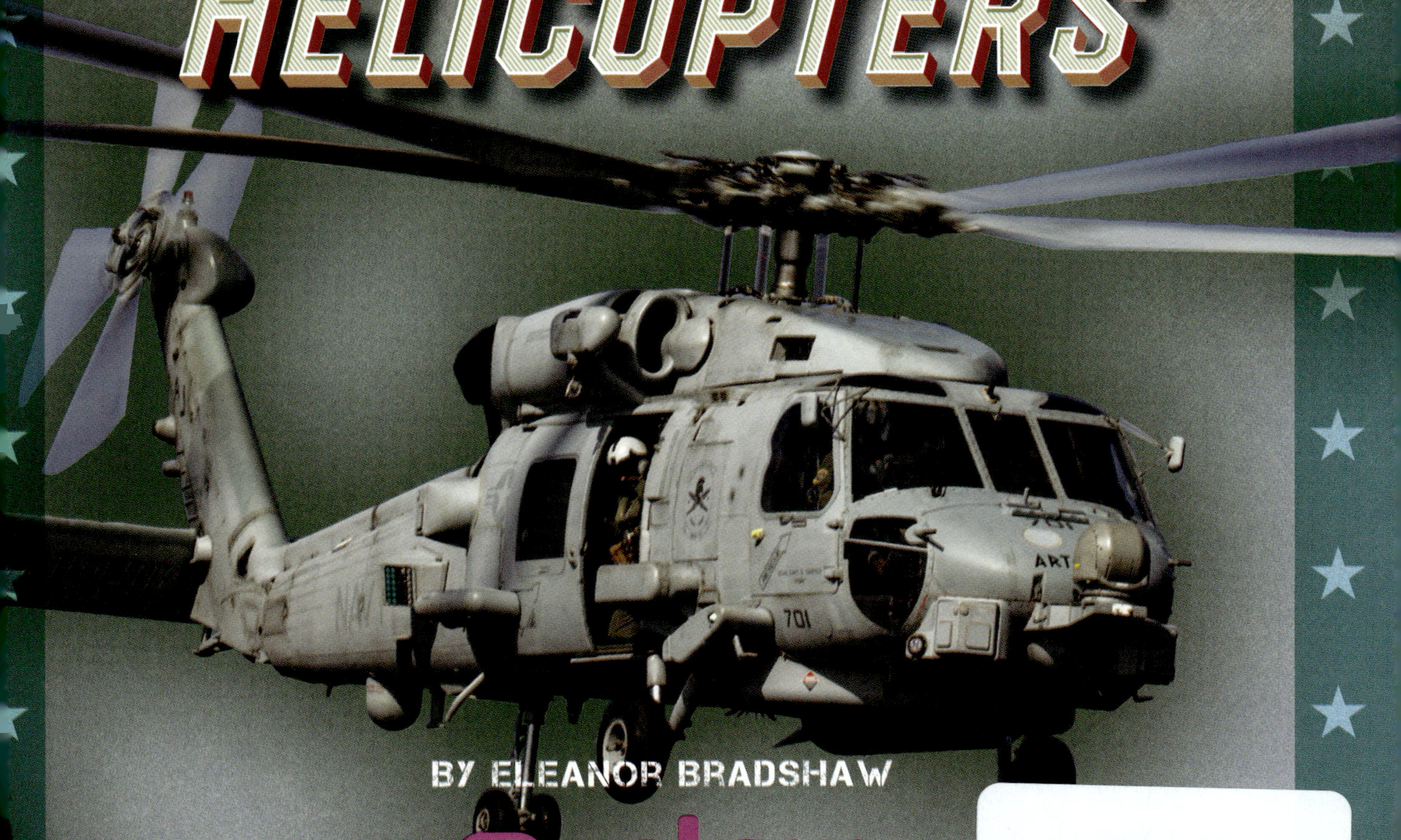

BY ELEANOR BRADSHAW

Enslow PUBLISHING

Please visit our website, www.enslow.com. For a free color catalog of all our high-quality books, call toll free 1-800-398-2504 or fax 1-877-980-4454.

Library of Congress Cataloging-in-Publication Data
Names: Bradshaw, Eleanor, author.
Title: Military helicopters / Eleanor Bradshaw.
Description: Buffalo, New York : Enslow Publishing, 2025. | Series: Amazing military machines | Includes index.
Identifiers: LCCN 2024001385 | ISBN 9781978539655 (library binding) | ISBN 9781978539648 (paperback) | ISBN 9781978539662 (ebook)
Subjects: LCSH: Military helicopters–United States–Juvenile literature.
Classification: LCC UG1233 .B724 2025 | DDC 623.74/6047209073–dc23/eng/20240206
LC record available at https://lccn.loc.gov/2024001385

Published in 2025 by
Enslow Publishing
2544 Clinton Street
Buffalo, NY 14224

Portions of this work were originally authored by Ryan Nagelhout and published as *Military Helicopters*. All new material in this edition is authored by Eleanor Bradshaw.

Designer: Andrea Davison-Bartolotta
Editor: Rachael Morlock

Photo credits: Cover U.S. Navy photo by Mass Communication Specialist 2nd Class Nolan Pennington/Navy.mil; series art (metal background texture) caesart/Shutterstock.com; series art (title font) TortugaStudios/Shutterstock.com; p. 5 (top) U.S. Army photo by Pfc. Ethan Scofield/DVIDS; p. 5 (bottom) U.S. Coast Guard photo by Petty Officer 1st Class Levi Read/DVIDS; p. 7 U.S. Navy photo by Mass Communication Specialist 3rd Class Caylen McCutcheon/DVIDS; p. 8 File:Igor Sikorsky B&W.jpg/Wikimedia Commons; p. 9 Pictorial Press Ltd/Alamy Stock Photo; p. 10 File:Sikorsky R-4 USAAF.jpg/Wikimedia Commons; p. 11 File:Igor Sikorsky in a U.S. Coast Guard HNS-1, 14 August 1944 (232-8).jpg/Wikimedia Commons; p. 13 Photo by Airman Allison Martin/Minot Air Force Base Public Affairs/DVIDS; p. 15 VanderWolf Images/Shutterstock.com; p. 17 courtesy of U.S. Department of Defense; p. 19 U.S. Marine Corps photo by 1st Lt. Mark Andries/DVIDS; p. 21 Photo by Victoria Falcon/Naval Air Station Patuxent River/DVIDS.

Printed in the United States of America

CPSIA compliance information: Batch #CS25ENS: For further information contact Enslow Publishing, at 1-800-398-2504.

CONTENTS

Boldface words appear in Words to Know.

FLYING MACHINES

Helicopters are amazing military machines with many different uses. Each branch of the U.S. military uses helicopters for different purposes. These machines are also known as rotorcraft. Instead of wings, they use sets of spinning blades, called rotors, to fly.

TAIL ROTOR
ROTOR

6558
U.S. COAST GUARD

Helicopters fly differently than airplanes. Airplanes can only move forward, but helicopters can move forward, backward, and sideways! **Pilots** change direction by turning the rotors. Unlike an airplane, a helicopter can take off and land **vertically**! It doesn't need much space.

Helicopters can drop off and pick up troops and **cargo** while hovering, or hanging in the air.

TAKING FLIGHT

The VS-300 was one of the first working helicopters. It was made by Igor Sikorsky and took flight on September 14, 1939. The helicopter was **tethered** during its short first flight. Sikorsky kept **developing** helicopters that could fly safely.

IGOR SIKORSKY

Igor Sikorsky, shown here, wanted the helicopter to be a tool for **rescuing** people from floods, fires, and unsafe places.

In 1942, the U.S. military purchased more than 100 Sikorsky R-4 helicopters. These machines had a large main rotor with three blades. They flew up to 75 miles (121 km) per hour and as high as 8,000 feet (2,438 m).

SIKORSKY R-4

The U.S. Army Air Forces, Navy, and Coast Guard used the Sikorsky R-4 helicopters.

WARTIME TOOLS

Helicopters became an important tool in the Korean War (1950–1953) for moving **wounded** troops to safety. In the Vietnam War (1954–1975), helicopters also carried troops and supplies. They could cross over mountains and through jungles to bring troops to battle.

The Bell UH-1 Iroquois, called the "Huey," moved troops and cargo during the Vietnam War. It could **attack** from the air.

ATTACK HELICOPTERS

Today, the U.S. Army uses many helicopters that are ready to attack. For example, the AH-64, or Apache, is loaded with **missiles**, rockets, and guns. It uses **radar** to find enemies. The Sikorsky UH-60 Black Hawk is another attack helicopter.

Black Hawk helicopters were first used in 1979. The U.S. Army still flies them today.

TRANSPORT HELICOPTERS

The main purpose of some military helicopters is to transport, or move, people and supplies. Chinook helicopters have room for 44 troops or 22 **stretchers**. Two big rotors help them fly and carry up to 11 tons (10 mt).

The Chinook can also transport items, such as cars and boats, using ropes.

MANY USES

Some helicopters have many purposes. The U.S. Navy, Marine Corps, and Air Force have all used the V-22, or Osprey. It has both rotors and wings. The Osprey can take off vertically with its rotors, then fly long distances with its wings.

The Osprey can land on ships like aircraft carriers, and its rotors fold in for storage.

NEW MACHINES

The U.S. military is building and testing new helicopters to use in place of the oldest aircraft. The newest helicopters are advanced machines with many uses. Helicopters can rescue people, transport the wounded, carry troops and supplies, and attack enemies.

The King Stallion CH-53K is a new helicopter for the U.S. Marine Corps. It moves marines from ship to shore.

WORDS TO KNOW

attack: To set upon with force and the intent to do harm.

cargo: Supplies or tools being moved.

develop: To create over time.

missile: A weapon sent out to hit something far away.

pilot: A person who flies an aircraft.

radar: A machine that uses radio waves to locate and identify objects.

rescue: To free a person or thing from unsafe conditions.

stretcher: A cot or bed used to carry a sick or injured person.

tether: To tie to a line that limits movement.

vertically: In a straight up and down manner.

wounded: Hurt by a weapon.

FOR MORE INFORMATION

BOOKS

Bolte, Mari. *Helicopters in Action*. Minneapolis, Minnesota: Lerner Publications, 2024.

Brody, Walt. *How Military Helicopters Work*. Minneapolis, Minnesota: Lerner Publications, 2020.

WEBSITES

CV-22 Osprey
www.af.mil/About-Us/Fact-Sheets/Display/Article/104531/cv-22-osprey/
This fact sheet from the U.S. Air Force offers detailed information about the Osprey helicopter.

Helicopters
www.military.com/equipment/helicopters
This website provides images of many of the types of helicopters used by the U.S. military.

Publisher's note to educators and parents: Our editors have carefully reviewed these websites to ensure that they are suitable for students. Many websites change frequently, however, and we cannot guarantee that a site's future contents will continue to meet our high standards of quality and educational value. Be advised that students should be closely supervised whenever they access the internet.

INDEX